THE PRACTICE OF
BUSINESS STATISTICS

COMPANION CHAPTER 13
TIME SERIES FORECASTING

David S. Moore
Purdue University

George P. McCabe
Purdue University

William M. Duckworth
Iowa State University

Stanley L. Sclove
University of Illinois

W. H. Freeman and Company
New York

Senior Acquisitions Editor:	**Patrick Farace**
Senior Developmental Editor:	**Terri Ward**
Associate Editor:	**Danielle Swearengin**
Media Editor:	**Brian Donnellan**
Marketing Manager:	**Jeffrey Rucker**
Head of Strategic Market Development:	**Clancy Marshall**
Project Editor:	**Mary Louise Byrd**
Cover and Text Design:	**Vicki Tomaselli**
Cover Illustration:	**Janet Hamlin**
Production Coordinator:	**Paul W. Rohloff**
Composition:	**Publication Services**
Manufacturing:	**RR Donnelley & Sons Company**

TI-83™ screens are used with permission of the publisher: ©1996, Texas Instruments Incorporated.

TI-83™ Graphics Calculator is a registered trademark of Texas Instruments Incorporated.

Minitab is a registered trademark of Minitab, Inc.

SAS© is a registered trademark of SAS Institute, Inc.

Microsoft© and Windows© are registered trademarks of the Microsoft Corporation in the USA and other countries.

Excel screen shots reprinted with permission from the Microsoft Corporation.

Cataloguing-in-Publication Data available from the Library of Congress

Library of Congress Control Number: 2002108463

Printed in the United States of America

Second Printing

TO THE INSTRUCTOR

NOW *YOU* HAVE THE CHOICE!

This is Companion Chapter 13 to *The Practice of Business Statistics (PBS)*. Please note that this chapter, along with any other Companion Chapters, can be bundled with the *PBS* Core book, which contains Chapters 1–11.

These other Companion Chapters, *in any combinations you wish,* are available for you to package with the *PBS* Core book.

TIME SERIES FORECASTING

The Holiday *Shipping* Season

Retail stores traditionally see a surge in sales beginning the day after the Thanksgiving holiday and continuing through December 24. This is called the "holiday shopping season" and is a very busy time for retail stores. Related to all this holiday shopping is a surge in the mailing of packages. Online shoppers have their purchases shipped. Many people mail gifts across the country or around the world. Package delivery companies refer to this as the "holiday *shipping* season." The surge in the number of packages to be delivered is huge—more than can be covered by employees working overtime.

One package delivery company, United Parcel Service (UPS), posted details of how the company handles the holiday shipping season on its Web site.[1] UPS estimates it will deliver 18 million packages during the "Peak Day" of the holiday shipping season. (For 2002, Peak Day was Thursday, December 19.) UPS's average number of packages delivered per day is 13.6 million. The extra 4.4 million packages of Peak Day are a 32% surge in UPS deliveries, and that is just one day of the holiday shipping season!

To handle the increase in business in 2002, UPS added about 60,000 seasonal employees to its usual workforce of 360,000 employees. (In past years, UPS has hired as many as 90,000 seasonal employees!) Without enough extra help, the holiday shipping season would be a disaster for UPS and its customers. How does UPS know how much extra help it will need? How does it estimate how many extra packages it will be asked to ship? UPS tracks the number of packages shipped over time. The regular ups and downs during the year are predictable, including the surge of the holiday shipping season. Time series methods identify such patterns and aid in forecasting future values.

Time Series Forecasting

Introduction

Business decisions often involve data tracked over time. Quarterly sales figures, annual health benefits costs, monthly product demand, daily stock prices, and changes in market share are all examples of *time series* data.

> **TIME SERIES**
>
> Measurements of a variable taken at regular intervals over time form a **time series**.

time plot

We've already encountered time series data in earlier chapters. Figure 1.7 (page 21) displayed a **time plot** of monthly orange prices from January 1991 to December 2000. Figure 1.8 (page 22) displayed a time plot of national unemployment rates from January 1990 to August 2001. We handle time series data with the same approach used in earlier chapters:

- Plot the data, then add numerical summaries.

- Identify overall patterns and deviations from those patterns.

- When the overall pattern is quite regular, use a compact mathematical model to describe it.

In this chapter, we will focus on the plots and calculations that are most helpful when describing time series data. We will learn to identify patterns common to time series data as well as the models that are commonly used to describe those patterns.

■
EXAMPLE 13.1   Monthly retail sales

The Census Bureau tracks retail sales using the Monthly Retail Trade Survey.[2] Retail establishments are categorized and tracked by their North American Industry Classification System (NAICS) code. NAICS code 452 corresponds to general merchandise stores (like Wal-Mart). Comparing sales data for a specific retail company with national sales data lets us see how the company is performing compared to its industry as a whole.

Figure 13.1 plots monthly retail sales of general merchandise stores (NAICS 452) beginning in January 1992 and ending in May 2002 (125 months). The plot reveals interesting characteristics of retail sales for these stores:

- Since January 1992, overall sales have gradually increased.

- A distinct pattern repeats itself approximately every 12 months: January sales are lower than the rest of the year; sales then increase but level off until October, when sales dramatically increase to a peak in December.

trend

The gradual increase in overall sales is an example of a **trend** in a time series. Embedded in the monthly ups and downs of sales figures, we also see a persistent pattern describing the general, long-run behavior of sales over the time period from January 1992 to May 2002.

seasonal variation

The repeating pattern evident in Figure 13.1 is an example of **seasonal variation** in a time series. Seasonal variation must be taken into account when

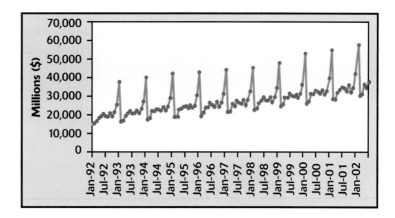

Figure 13.1 Time plot of U.S. retail sales of general merchandise stores for each month from January 1992 to May 2002, for Example 13.1.

making a short-run prediction of future sales. It would be unwise to ignore the peaks in December and the troughs in January when predicting sales for specific months.

forecast

"Trend" and "seasonal variation" were defined in Chapter 1 (page 20). A prediction of a future value of a time series is called a **forecast**. This chapter focuses on exploring time series data with the eventual goal of forecasting the variable of interest.

APPLY YOUR KNOWLEDGE

13.1   **JCPenney sales.** Table 13.1 contains retail sales for JCPenney in millions of dollars. The data are quarterly, beginning with the first quarter of 1996 and ending with the fourth quarter of 2001.[3]

TABLE 13.1	Quarterly retail sales for JCPenney, 1996–2001 (millions of dollars)		
Year-quarter	Sales	Year-quarter	Sales
1996-1st	4452	1999-1st	7339
1996-2nd	4507	1999-2nd	7104
1996-3rd	5537	1999-3rd	7639
1996-4th	8157	1999-4th	9661
1997-1st	6481	2000-1st	7528
1997-2nd	6420	2000-2nd	7207
1997-3rd	7208	2000-3rd	7538
1997-4th	9509	2000-4th	9573
1998-1st	6755	2001-1st	7522
1998-2nd	6483	2001-2nd	7211
1998-3rd	7129	2001-3rd	7729
1998-4th	9072	2001-4th	9542

(a) Before plotting these data, inspect the values in the table. Do you see any interesting features of JCPenney quarterly sales?

(b) Now, make a time plot of the data. Be sure to connect the points in your plot to highlight patterns.

(c) Is there an obvious trend in JCPenney quarterly sales? If so, is the trend positive or negative?

(d) Is there an obvious repeating pattern in the data? If so, clearly describe the repeating pattern.

13.1 Trends and Seasons

In the modern business environment, large and small companies alike depend on long-run forecasts of numerous variables to guide their long-run business plans. A retail chain uses population growth forecasts to help decide where to locate future stores. A manufacturer of car alarm systems looks at auto theft growth rates in key areas of the country to estimate demand for their products. Some companies provide one-year-ahead forecasts of key quantities in their annual reports, so that a 2001 annual report will contain forecasts for how the company will perform in 2002. The first step toward a reliable forecast is to identify any trend in the time series of interest.

Identifying trends

EXAMPLE 13.2

Monthly retail sales

Consider the monthly retail sales data from Example 13.1. Since January 1992, overall sales have gradually increased. We can use the ideas of simple linear regression (Chapter 10) and software to estimate this upward trend. From the Excel regression output below, we estimate the linear trend to be

$$\widehat{SALES} = 18{,}736 + 145.5x$$

where x is the number of months elapsed beginning with the first month of the time series; that is, $x = 1$ corresponds to January 1992, $x = 2$ corresponds to February 1992, etc. Sales are measured in millions of dollars.

◇	A	B	C	D	E
1	SUMMARY OUTPUT				
2					
3	*Regression Statistics*				
4	Multiple R	0.655338351			
5	R Square	0.429468354			
6	Adjusted R Square	0.424829885			
7	Standard Error	6101.522201			
8	Observations	125			
9					
10		*Coefficients*	*Standard Error*	*t Stat*	*P-value*
11	Intercept	18736.49742	1098.055282	17.0633462	3.1962E-34
12	X	145.5311521	15.12438406	9.6222862	1.11261E-16

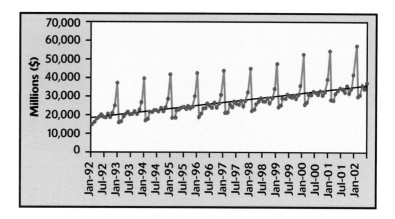

Figure 13.2 Trend line (*black*) fitted to U.S. retail sales of general merchandise stores, for Example 13.2.

Figure 13.2 displays the time series plot of Figure 13.1 with the trend line superimposed. The equation of the line provides a mathematical model for the observed trend in sales. The estimated slope of 145.5 indicates that retail sales at general merchandise stores increased an average of $145.5 million per month from January 1992 to May 2002.

Notice how the trend line in Figure 13.2 ignores the seasonal variation in the retail sales time series. Because of this, using the equation above to forecast sales for, say, December 2002 will likely result in a gross underestimate since the line underestimates sales for all the Decembers in the data set. We will need to take the month-to-month pattern into account if we wish to accurately forecast sales in a specific month.

In Chapter 11, we used regression techniques to fit a variety of models to data. For example, we used a polynomial model to predict the prices of houses in Example 11.14 (page 672). Trends in time series data may also be best described by a curved model like a polynomial. The techniques of Chapter 11 help us fit such trend curves to time series data. Case 13.1 *exponential trend* illustrates another type of curved relationship—**exponential trend**.

CASE 13.1

SELLING DVD PLAYERS

The popularity of the DVD format has exploded in the relatively short period of time since its introduction in March 1997. At the end of June 2002, nearly 33 million DVD players had been sold in the United States, with over 18,000 titles available in the DVD format. The Consumer Electronics Association (CEA) tracks monthly sales of DVD players. The data are provided in the file *ca13_001.dat*.[4] Figure 13.3 plots the number of units sold over time. The plot includes sales data from April 1997 to June 2002 (63 months).

Let your eye follow the sales numbers from April to April. DVD player sales increased very little from April 1997 to April 1998. Notice how

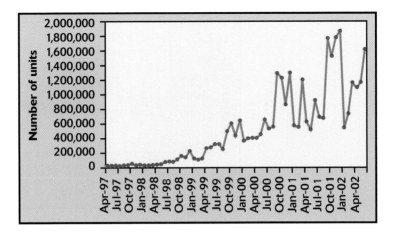

Figure 13.3 Time plot of number of DVD players sold for each month from April 1997 to June 2002, for Case 13.1.

the April-to-April sales increases have grown in magnitude over the years with April 2001 to April 2002 showing the largest increase. This pattern of increasing growth is an example of an *exponential trend*. Figure 13.4 shows the DVD player sales data with an exponential trend superimposed. Statistical software estimates the exponential trend to be

$$\widehat{\text{UNITS}} = 29{,}523.65e^{0.068941x}$$

where x is the number of months elapsed beginning with the first month of the time series; that is, $x = 1$ corresponds to April 1997, $x = 2$ corresponds to May 1997, etc. (The mathematical constant e was introduced in Chapter 5 on page 336.)

We can use our trend model to forecast the number of DVD players to be sold in a future month. July 2002 corresponds to $x = 64$, and we get

$$\widehat{\text{UNITS}} = 29{,}523.65e^{0.068941 \times 64} = 2{,}434{,}303$$

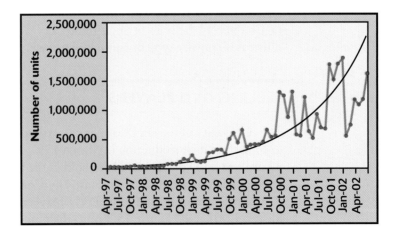

Figure 13.4 Exponential trend (*black*) fitted to the number of DVD players sold, for Case 13.1.

Our forecasted value ignores the monthly ups and downs in the time series. In fact, if you look closely at the data, you will see that July sales have always been less than June sales (except in 1998). Our trend model ignores such information about the time series. Our forecast of over 2.4 million DVD players to be sold in July 2002 is greater than the 1.6 million DVD players sold in June 2002.

Remember, a trend equation may be a good description of the long-run behavior of the data, but we need to account for short-run phenomena (like seasonal variation) to improve the accuracy of our forecasts.

APPLY YOUR KNOWLEDGE

13.2   **Number of Macs shipped.** Table 13.2 displays the time series of the number of Macintosh computers shipped in each of eight consecutive fiscal quarters.[5] We usually want a time series longer than just eight time periods, but this short series will give you a chance to do some calculations by hand. Hand calculations take the mystery out of what computers do so quickly and efficiently for us.

(a) Make a time plot of these data.

(b) Using the following summary information, calculate the least-squares regression line for predicting the number of Macs shipped (in thousands of units). The variable Time simply takes on the values 1, 2, 3, ..., 8 in time order.

Variable	Mean	Std. Dev.	Correlation
Time	4.5	2.44949	0.3095
Macs shipped	773.5	62.56197	

(c) Sketch the least-squares line on your time plot from part (a). Does the linear model appear to fit these data well?

TABLE 13.2	Macintosh computers shipped per fiscal quarter (thousands of units)
Fiscal year-quarter	**Units shipped**
2001-1st	659
2001-2nd	751
2001-3rd	827
2001-4th	850
2002-1st	746
2002-2nd	813
2002-3rd	808
2002-4th	734

13.3 **JCPenney sales.** In Exercise 13.1, you took a first look at the data in Table 13.1. Use statistical software to further investigate the JCPenney sales data.

(a) Find the least-squares line for the sales data. Use 1, 2, 3, ... as the values for the explanatory variable, with $x = 1$ corresponding to the first quarter of 1996, $x = 2$ corresponding to the second quarter of 1996, etc.

(b) The intercept is a prediction of sales for what quarter?

(c) Interpret the slope in the context of JCPenney quarterly sales.

Seasonal patterns

Variables of economic interest are often tied to other events that repeat with regular frequency over time. Agriculture-related variables will vary with the growing and harvesting seasons. In Example 1.7 (page 20), we saw that the prices of oranges tend to be lowest in January and February and highest in August and September. Sales data may be linked to events like regular changes in the weather, the start of the school year, and the celebration of certain holidays. The monthly retail sales data of Example 13.1 (page 13-4) exhibit a strong pattern that repeats every 12 months, with sales peaking in December. To improve the accuracy of our forecasts we need to account for seasonal variation in our time series.

Using indicator variables

Indicator variables were introduced in Chapter 11 (page 675). We can use indicator variables to add the seasonal pattern in a time series to a trend model. Let's look at the details for the monthly retail sales data of Examples 13.1 and 13.2.

EXAMPLE 13.3 **Monthly retail sales**

The seasonal pattern in the sales data seems to repeat every 12 months, so we begin by creating $12 - 1 = 11$ indicator variables.

$$X1 = \begin{cases} 1 & \text{if the month is January} \\ 0 & \text{otherwise} \end{cases}$$

$$X2 = \begin{cases} 1 & \text{if the month is February} \\ 0 & \text{otherwise} \end{cases}$$

$$\vdots$$

$$X11 = \begin{cases} 1 & \text{if the month is November} \\ 0 & \text{otherwise} \end{cases}$$

December data are indicated when all 11 indicator variables are 0.

We can extend the trend model estimated in Example 13.2 with these indicator variables. The new model will capture the seasonal pattern in the time series.

$$\text{SALES} = \beta_0 + \beta_1 x + \beta_2 X1 + \cdots + \beta_{12} X11 + \epsilon$$

Fitting this multiple regression model to our data, we get the following output:

◇	A	B	C	D	E	F	G
1	SUMMARY OUTPUT						
2							
3	*Regression Statistics*						
4	Multiple R	0.993493346					
5	R Square	0.987029029					
6	Adjusted R Square	0.985639282					
7	Standard Error	964.1133873					
8	Observations	125					
9							
10	ANOVA						
11		*df*	*SS*	*MS*	*F*	*Significance F*	
12	Regression	12	7921942577	660161881.4	710.2221576	1.1832E-99	
13	Residual	112	104105637.8	929514.6235			
14	Total	124	8026048215				
15							
16		*Coefficients*	*Standard Error*	*t Stat*	*P-value*	*Lower 95%*	*Upper 95%*
17	Intercept	37472.69024	343.3506365	109.1382577	1.4241E-115	36792.38539	38152.9951
18	X	140.1304508	2.392701887	58.5657794	7.61806E-86	135.3896217	144.87128
19	X1	−24276.1932	421.4213062	−57.60551933	4.57481E-85	−25111.18498	−23441.20142
20	X2	−23748.68729	421.3601691	−56.36196544	4.86201E-84	−24583.55793	−22913.81664
21	X3	−20271.18137	421.3126119	−48.1143474	1.19904E-76	−21105.95779	−19436.40496
22	X4	−20250.49364	421.2786392	−48.06912043	1.32582E-76	−21085.20275	−19415.78454
23	X5	−18517.98773	421.2582543	−43.95875343	1.79155E-72	−19352.65645	−17683.31902
24	X6	−19574.51729	431.403553	−45.37402894	6.21054E-74	−20429.28761	−18719.74698
25	X7	−20323.64775	431.330558	−47.11849733	1.1186E-75	−21178.27343	−19469.02206
26	X8	−18626.8782	431.2708257	−43.1906753	1.15687E-71	−19481.38553	−17772.37087
27	X9	−20878.10865	431.2243614	−48.41588397	6.14847E-77	−21732.52391	−20023.69338
28	X10	−18933.0391	431.1911697	−43.9086893	2.02136E-72	−19787.3886	−18078.6896
29	X11	−13842.16955	431.1712534	−32.10364662	2.57955E-58	−14696.47959	−12987.85951

Figure 13.5 displays the time series plot of Figure 13.1 with the trend-and-season model superimposed. You can see the dramatic improvement over the trend-only model by comparing Figure 13.2 with Figure 13.5. The improved model fit is reflected in the R^2-values for the two models: $R^2 = 0.987$ for the trend-and-season model and $R^2 = 0.429$ for the trend-only model.

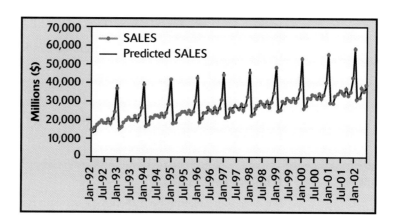

Figure 13.5 Trend-and-season model (*black*) fitted to U.S. retail sales of general merchandise stores, for Example 13.3.

APPLY YOUR KNOWLEDGE

13.4 **Number of Macs shipped.** In Exercise 13.2, you made a time plot of the data in Table 13.2. With only eight quarters, a strong quarterly pattern is hard to detect. Add indicator variables for first, second, and third quarters to the linear trend model fitted in Exercise 13.2. Call these indicator variables $X1$, $X2$, and $X3$, respectively. Use statistical software to fit this multiple regression model.

(a) Write down the estimated trend-and-season model.

(b) Explain why no indicator variable is needed for fourth quarters.

(c) What does the ANOVA F test indicate about this model?

13.5 **JCPenney sales.** In Exercise 13.1, you made a time plot of the data in Table 13.1. Sales seem to follow a pattern of ups and downs that repeats every four quarters. Add indicator variables for first, second, and third quarters to the linear trend model fitted in Exercise 13.3. Call these indicator variables $X1$, $X2$, and $X3$, respectively.

(a) Write down the estimated trend-and-season model.

(b) Explain why no indicator variable is needed for fourth quarters.

(c) Does the intercept still predict sales for a specific quarter? If so, what quarter? Compare the estimated intercept of this model with that of the trend-only model (see Exercise 13.3(b)). Given the pattern of seasonal variation, which appears to be the better estimate?

Using seasonality factors

Using indicator variables to incorporate seasonality into a trend model views the model as a trend component plus a seasonal component:

$$y = \text{TREND} + \text{SEASON}$$

You can see this in Example 13.3, where we *added* the indicator variables to the trend model. A second approach to accounting for seasonal variation is to calculate an "adjustment" factor for each season. The trend is adjusted each particular season by *multiplying* it by the appropriate **seasonality factor**. One seasonality factor is calculated for each season observed in the data. Using seasonality factors views the model as a trend component times a seasonal component.

seasonality factor

$$y = \text{TREND} \times \text{SEASON}$$

This view of our model suggests that we can identify the seasonal component by calculating

$$\frac{y}{\text{TREND}} = \text{SEASON}$$

Let's see how this works in practice using the monthly retail sales data of Example 13.1.

EXAMPLE 13.4 **Monthly retail sales**

We have been treating each month as a season for the monthly retail sales data, so we will need to calculate 12 seasonality factors. In Example 13.2 (page 13-6), the trend component was estimated to be

$$\widehat{\text{SALES}} = 18{,}736 + 145.5x$$

For each of the 125 months in our time series, we calculate the ratio of actual sales divided by predicted sales, $\widehat{\text{SALES}}$. We then average these ratios by month. That is, we compute the average for all the January ratios, then the average for all the February ratios, and so on. These averages become our 12 seasonality factors. The arithmetic needed is straightforward but tedious given that we have 125 months of data. The table below displays the seasonality factors that result.

Month	Seasonality factor	Month	Seasonality factor
January	0.784	July	0.931
February	0.805	August	0.994
March	0.931	September	0.913
April	0.935	October	0.985
May	0.995	November	1.166
June	0.958	December	1.662

Our trend-and-season model is then

$$\widehat{\text{SALES}} = (18{,}736 + 145.5x) \times \text{SF}$$

where SF is the seasonality factor for the appropriate month corresponding to the value of x.

The seasonality factors are a snapshot of the typical ups and downs over the course of a year. The factors are used to adjust the trend model to account for the differences from season to season (month to month in this example). The factors themselves have an interpretation if their average is 1—our factors have an average of 1.005, which is very close to 1.[*] If you think of the factors as percents, then the factors show how each month's sales compare to average monthly sales for all 12 months. For example, December's seasonality factor is 1.662, or 166.2%, indicating that December sales are typically 66.2% above the average for all 12 months. January's seasonality factor is 0.784, or 78.4%, indicating that January sales are typically 21.6% below the annual average. Figure 13.6 plots the seasonality factors in order from January to December. The reference line marked at 1.0 (100%) is a visual aid for interpreting the factors compared to the overall average monthly sales. Notice how the seasonality factors mimic the pattern that is repeated every 12 months in Figure 13.1—dividing the sales data by the estimated trend has isolated the seasonal pattern in the time series.

---

[*] If the factors did not have an average close to 1, we could adjust them so that they did average to 1 by dividing each factor by the total of all 12 factors.

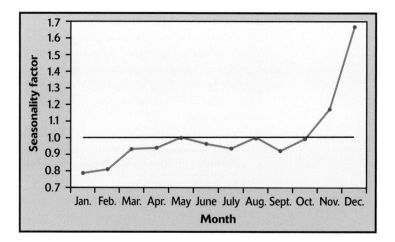

Figure 13.6 Seasonality factors for the U.S. retail sales of general merchandise stores plotted against month, for Example 13.4.

EXAMPLE 13.5

CASE 13.1

Next month's DVD player sales

Identifying an appropriate model to capture trend and, if present, seasonal variation is an important first step in forecasting future values of a time series. In Case 13.1 (page 13-7), we forecasted the number of DVD players that would sell in July 2002 based on data ending in June 2002. Our forecast used a trend-only model. Let's forecast July 2002 DVD player sales using a trend-and-season model.

The trend-only forecast for DVD player sales in July 2002 ($x = 64$) was calculated as

$$\widehat{\text{UNITS}} = 29{,}523.65 e^{0.068941 \times 64} = 2{,}434{,}303$$

If we calculate the seasonality factors for the DVD player sales data using each month as a season, we get the factors displayed in the following table.

Month	Seasonality factor	Month	Seasonality factor
January	0.693	July	0.871
February	0.661	August	0.820
March	0.836	September	1.418
April	0.848	October	1.506
May	0.781	November	1.123
June	0.989	December	1.454

Our July trend-only prediction must be multiplied by the July seasonality factor 0.871 to take seasonal variation into account in our forecast:

$$\widehat{\text{UNITS}} = (29{,}523.65 e^{0.068941 \times 64}) \times 0.871 = 2{,}120{,}278$$

Taking seasonal variation into account, we forecast July DVD player sales to be 12.9% below what is forecasted by the exponential trend alone. This forecast is less than the trend-only forecast, but it may still be too high given our observation that July sales are always below June sales (except in 1998), and June 2002 sales were only 1,617,098 units. The seasonal pattern is not as strong as the monthly retail sales data of Example 13.1, so we may need more than a trend-and-season model to accurately forecast monthly DVD player sales.

▪ ▪ ▪

Trend-only and trend-and-season models using indicator variables are both regression models like those used in Chapters 10 and 11. Forecasting with these models uses the same techniques as the regression-based predictions in Chapters 10 and 11. However, the prediction intervals presented in those chapters cannot typically be used with time series data. The regression models used in this section will usually have residuals that indicate that our regression assumptions—particularly the assumption of independent deviations ϵ—are not appropriate for our time series data.

APPLY YOUR KNOWLEDGE

13.6 **Number of Macs shipped.** In Exercise 13.2, you fitted a linear trend-only model to the time series of Macs shipped. Starting with this trend model, incorporate seasonality factors for each quarter. With only eight quarters of data, these calculations can be done by hand (or computer).

(a) Calculate the seasonality factor for each quarter.

(b) Average the four seasonality factors. Is this average close to 1? If so, interpret the seasonality factor for first quarters.

(c) Make a scatterplot of seasonality factor versus quarter with the seasonality factors on the vertical axis and the quarters on the horizontal axis. Connect the points to see the general pattern of seasonal variation. Also, draw a horizontal line at the average of the four seasonality factors.

13.7 **JCPenney sales.** In Exercise 13.3, you fitted a linear trend-only model to the JCPenney sales data. Starting with this trend model, we want to incorporate seasonality factors to account for the pattern that repeats every four quarters.

(a) Calculate the seasonality factor for each quarter.

(b) Average the four seasonality factors. Is this average close to 1? If so, interpret the seasonality factor for fourth quarters.

(c) Make a scatterplot of seasonality factor versus quarter with the seasonality factors on the vertical axis and the quarters on the horizontal axis. Connect the points to see the general pattern of seasonal variation. Also, draw a horizontal line at the average of the four seasonality factors.

Seasonally adjusted data

Many economic time series are seasonally adjusted to make the overall trend in the numbers more apparent. Government agencies will often release both versions of a time series, so be careful to notice whether you are analyzing seasonally adjusted data or not when using government sources. A seasonally adjusted time series has had each value divided by the seasonality factor corrresponding to the appropriate month.

EXAMPLE 13.6

Seasonally adjusted monthly retail sales

To calculate seasonally adjusted monthly retail sales, we simply divide each actual sales value by the appropriate seasonality factor calculated in Example 13.4.

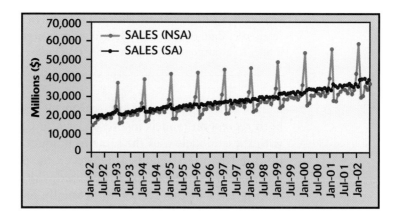

Figure 13.7 Sales (NSA) is the original U.S. retail sales of general merchandise stores (*purple*). Sales (SA) is the seasonally adjusted sales data (*black*), for Example 13.6.

Figure 13.7 displays the original time series and the seasonally adjusted time series. The seasonally adjusted time series, SALES (SA), does not show the regular ups and downs of the data that are not seasonally adjusted, SALES (NSA). The seasonally adjusted time series shows the overall trend and some random variation. Seasonally adjusted data are easier to interpret because the technique flattens the regular peaks and valleys that we should expect in our data and makes it easier to identify unusual peaks and/or valleys in the data.

There are many different approaches to seasonal adjustment. The indicator variable model of Example 13.3 can be used, although we won't go through the details. We could also use a trend model that is more complicated than a line. Specialized time series models can also be used to seasonally adjust a time series. Different approaches will lead to different estimates and forecasts; however, the differences will not usually be great.

APPLY YOUR KNOWLEDGE

13.8 Seasonally adjusted DVD player data

(a) Make a scatterplot of the seasonality factors in Example 13.5 with the seasonality factors on the vertical axis and the months (in time order) on the horizontal axis. Connect the points and describe the overall pattern of seasonal variation in the DVD player sales data.

(b) Calculate the seasonally adjusted value of the time series for June 2002. Do the calculation by hand.

13.9 Seasonally adjusted DVD player data

(a) Using statistical software, calculate the seasonally adjusted time series of the DVD player sales data. Make a time plot of the original DVD player time series with the seasonally adjusted time series superimposed (see Figure 13.7 for an example).

(b) Did seasonally adjusting the DVD player sales data smooth the time series to the degree that seasonally adjusting the sales data in Figure 13.7

did? What does this imply about the strength of the seasonal pattern in these two time series?

Looking for autocorrelation

Using a regression model with time series data as we've done in this section will often result in residuals that indicate that our regression assumption of independent deviations is not appropriate. The current value of a time series may be heavily influenced by past values of the same series. A simple form of this dependence on past values is known as *autocorrelation*. In Chapter 2, we defined correlation as a measure of the strength of the linear relationship between two variables x and y (page 103). Similarly, autocorrelation measures the strength of the linear relationship between successive values of the same variable.

AUTOCORRELATION

The correlation between successive values y_{t-1} and y_t of a time series is called **first-order autocorrelation** or, simply, **autocorrelation.**

The residuals from a regression model that uses time as an explanatory variable should be examined for signs of autocorrelation. If successive residuals tend to be both negative or both positive, then the residuals will have positive autocorrelation. If successive residuals tend to be of opposite sign, then the residuals will have negative autocorrelation. In either case, we have evidence that our regression assumption of independent deviations ϵ is inappropriate.

Let's examine the residuals that resulted from fitting an exponential trend to the DVD player sales data from Case 13.1.

EXAMPLE 13.7

CASE 13.1

DVD player sales

Figure 13.8 plots the residuals from our exponential trend-only model for DVD player sales. Notice the pattern:

- Starting on the left side of the plot, negative residuals tend to be followed by another negative residual.

- In the middle of the plot, positive residuals tend to be followed by another positive residual.

- On the right side of the plot, negative residuals tend to be followed by another negative residual.

This pattern indicates positive autocorrelation among the residuals.

An alternative plot for detecting autocorrelation is given in Figure 13.9. Because autocorrelation measures the strength of the linear relationship between successive values in a series, plotting pairs of successive values will

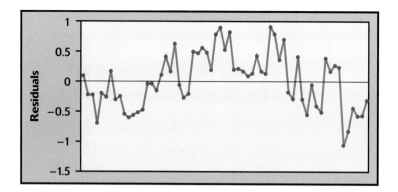

Figure 13.8 Time plot of residuals from the exponential trend model for DVD player sales, for Example 13.7.

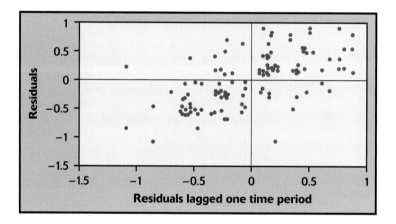

Figure 13.9 Lagged residual plot of residuals from the exponential trend model for DVD player sales, for Example 13.8.

show a strong linear pattern when significant autocorrelation exists. Figure 13.9 plots points of the form (e_{t-1}, e_t) where e_t is the residual for time period t. A scatterplot of the "lagged" residuals e_{t-1} against the residuals e_t will show a strong positive association for a series that has positive autocorrelation and a strong negative association for a series that has negative autocorrelation. If the points in a *lagged residual plot* (like Figure 13.9) are equally dispersed among the four quadrants of the scatterplot, then we do not have visual evidence of autocorrelation in our residuals.

LAGGED RESIDUAL PLOT

A **lagged residual plot** is a scatterplot of residuals e_t against the same residuals lagged one time period; that is, we plot the points $(e_1, e_2), (e_2, e_3), \ldots, (e_{n-1}, e_n)$. The plot is used to detect autocorrelation in residuals from a time series. A positive association in the points (e_{t-1}, e_t) indicates positive autocorrelation, while a negative association indicates negative autocorrelation.

EXAMPLE 13.8

DVD player sales

The linear pattern with positive slope in Figure 13.9 indicates that our residuals may have positive autocorrelation. The correlation for Figure 13.9 is 0.6165. This is approximately the autocorrelation for the residuals lagged one time period. Statistical software calculates the autocorrelation to be 0.6135 using a modified version of our correlation formula from Chapter 2.

■—■—■—■

Durbin-Watson test

The positive autocorrelation of 0.6135 is another bit of evidence that our residuals are dependent, but is 0.6135 a large enough autocorrelation to give us confidence in our conclusion of dependence? To answer this question conclusively, we would use a statistical test for autocorrelation—the **Durbin-Watson test.** The Durbin-Watson test statistic is provided with the multiple regression output of many statistical software packages. You can consult a multiple regression text for the details of this test.[6]

Several methods are proposed for handling autocorrelation in a time series. You might add lagged values of your residuals as explanatory variables to a trend-only model. Other approaches suggest clever transformations of the variables in the model to adjust for the autocorrelation. Another approach is to fit a model that uses only past values of the time series itself for forecasting rather than a model that uses other explanatory variables for forecasting the time series variable.

APPLY YOUR KNOWLEDGE

13.10   **DVD player sales.** In Examples 13.7 and 13.8, we found autocorrelation in the residuals from the exponential trend-only model for DVD player sales. The residuals from a trend-and-season model may not exhibit the same pattern of autocorrelation.

(a)   Calculate the predicted number of units sold using the trend-and-season model based on the seasonality factors given in Example 13.5. That is, calculate

$$\widehat{\text{UNITS}} = (29{,}523.65e^{0.068941x}) \times \text{SF}$$

for each month in the time series. Be sure to use the appropriate seasonality factor (SF) for each month.

(b)   Calculate the residuals for the model in part (a). That is, calculate

$$e = \text{UNITS} - \widehat{\text{UNITS}}$$

for each month in the time series.

(c)   Make a time plot of the residuals. Do we have visual evidence of autocorrelation in this plot? Describe the evidence.

(d)   Make a lagged residual plot and calculate the correlation between successive residuals e_{t-1} and e_t. Do we have evidence of autocorrelation? Describe the evidence.

SECTION 13.1 SUMMARY

■ Data collected at regular time intervals are called a **time series.** Time series often display a long-run **trend.** Some time series also display a strong, repeating **seasonal pattern.**

■ Regression methods can be used to model the trend and seasonal variation in a time series. Indicator variables can be used to model the seasons in a time series. Alternatively, **seasonality factors** can be used to adjust trend-only models for seasonal effects.

■ **Seasonally adjusted** data have been divided by seasonality factors to remove the effect of seasonal variation on the time series. Government agencies typically release seasonally adjusted data for economic time series.

■ Successive residuals may be linearly related. A **lagged residual plot** is a scatterplot of residuals plotted against the residuals lagged by one time period. If residuals exhibit **autocorrelation**, then our usual regression assumptions are not appropriate; in particular, the deviations cannot be assumed to be independent. In this case, a model based on past values of the time series should be considered.

SECTION 13.1 EXERCISES

13.11 **JCPenney sales.** A linear trend-only model was fitted to the JCPenney sales data in Exercise 13.3. Use this model to answer the following:
  (a) On a time plot of the sales data, draw the least-squares line. Comment on any pattern of over- or underprediction if we were to use this trend-only model for predicting sales.
  (b) Using the equation of the least-squares line, forecast sales for the first quarter of 2002 and for the fourth quarter of 2002.
  (c) Which forecast in part (b) do you believe will be more accurate when compared to actual JCPenney sales? Why?

13.12 **JCPenney sales.** A trend-and-season model with indicator variables was fitted to the JCPenney sales data in Exercise 13.5.
  (a) Using the equation of the trend-and-season model, forecast sales for the first quarter of 2002 and for the fourth quarter of 2002.
  (b) Compare your forecasts with the same forecasts based on the trend-only model of Exercise 13.11.

13.13 **JCPenney sales.** A trend-and-season model using seasonality factors was calculated in Exercise 13.7.
  (a) Using the linear trend-only model and the seasonality factors, forecast sales for the first quarter of 2002 and for the fourth quarter of 2002.
  (b) Compare your forecasts with the same forecasts based on the trend-only model of Exercise 13.11.
  (c) Compare your forecasts with the same forecasts based on the trend-and-season model of Exercise 13.12.

13.14 **JCPenney sales.** The model in part (a) of Exercise 13.5 can be viewed as four models—one for each quarter—by setting the indicator variables to 0 or 1 for each of the four quarters (our seasons for this data set).
  (a) Write down the four models (one for each quarter) that are derived from the trend-and-season model in Exercise 13.5.
  (b) Each of the models in part (a) is linear. What do you notice about the four slopes?
  (c) On a time plot of the sales data, sketch each of the four lines. What geometric property do the lines possess?

13.15 **Comparing models for JCPenney sales.** Compare the trend-only model of Exercise 13.3 with the trend-and-season model of Exercise 13.5.

(a) Report the value of R^2 for each model and comment on the difference.

(b) Report the value of s for each model and comment on the difference.

(c) Make a time plot of the original JCPenney sales figures. On this plot, overlay both the trend-only predictions and the trend-and-season predictions.

(d) Taking into account parts (a), (b), and (c), is the trend-and-season model a big improvement over the trend-only model?

13.16 **Seasonally adjusted JCPenney sales.** In Exercise 13.7, you calculated seasonality factors for the JCPenney quarterly sales data. Using these factors, complete the following:

(a) Calculate the seasonally adjusted value of the time series for the fourth quarter of 2001. Do the calculation by hand.

(b) Using statistical software, calculate the seasonally adjusted JCPenney sales time series. Make a time plot of the original sales data with the seasonally adjusted sales data superimposed (see Figure 13.7 on page 13-16 for an example).

(c) Did seasonally adjusting the JCPenney sales data smooth the time series to the degree that seasonally adjusting the sales data in Figure 13.7 did? What does this imply about the strength of the seasonal pattern in these two time series?

13.17 **Autocorrelation in the JCPenney time series.** In Exercise 13.3, a linear trend-only model was fitted to the JCPenney sales data. Using the residuals from this model, look for evidence of autocorrelation.

(a) Make a time plot of the residuals. Describe any pattern you see in this plot.

(b) Make a lagged residual plot and calculate the correlation between successive residuals e_{t-1} and e_t. Do we have evidence of autocorrelation?

13.18 **Autocorrelation in the JCPenney time series.** Repeat the previous exercise using the residuals from the trend-and-season model of Exercise 13.5.

lagged time series plot **13.19** **Lagged time series plot.** Similar to the lagged residual plot, a **lagged time series plot** is a scatterplot with y_t on the vertical axis and y_{t-1} on the horizontal axis. Figure 13.10 is a lagged time series plot of the monthly retail sales data introduced in Example 13.1.

(a) Figure 13.10 consists of three distinct groups of points. One large group contains most of the points, and two smaller groups consist of 10 points each. Given what we know about when the ups and downs occur with this time series, we can identify the two smaller groups of points. The group of 10 points crossing the lower-right corner are the points with January sales as the y coordinate. The other group of 10 points has which month as the y coordinate?

(b) The correlation between successive months' sales is 0.4573. If we exclude the two groups of 10 points identified in part (a), the correlation is 0.9206. Does this suggest strong autocorrelation in the time series?

(c) If you looked at the correlation between successive values of the seasonally adjusted time series, would you expect the correlation to be closer to 0.4573 or 0.9206? Explain your response.

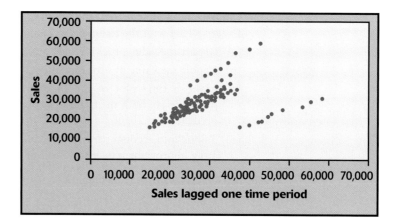

Figure 13.10 Lagged time series plot of U.S. retail sales of general merchandise stores, for Exercise 13.19.

13.2 Time Series Models

The previous section applied regression methods from Chapters 10 and 11 to time series data. A time period variable was used as the explanatory variable, and the time series was the response variable. Models like these may not satisfy our usual regression assumptions—particularly the independence assumption. Specific models have been developed to aid in the analysis of time series data when usual regression methods are not appropriate.

> **TIME SERIES MODELS**
>
> **Time series models** use past values of the time series to predict future values of the time series.

In the language of regression, a future time period's value is our "response" while one or more past time periods' values (of the same variable) are the explanatory variable(s). This approach is different from using the values of one variable x to predict the values of a second variable y. A time series model makes forecasts based on past values of the time series itself. For example, we might forecast next month's DVD player sales to be some multiple of this month's DVD player sales. Instead, we might use average DVD player sales for the last three months to predict DVD player sales this month. A wide variety of time series models exist to implement forecasting strategies like these as well as more elaborate forecasting strategies. This section will introduce a few of the most common time series models and illustrate how to use them for forecasting future values of a time series.

Autoregressive models

Can yesterday's stock price help predict today's stock price? Can last quarter's sales be used to predict this quarter's sales? Sometimes the best explanatory variables are simply past values of the response variable.

Autoregressive time series models take advantage of the linear relationship between successive values of a time series to predict future values of the series.

FIRST-ORDER AUTOREGRESSION MODEL

The **first-order autoregression model** specifies a linear relationship between successive values of the time series. The shorthand for this model is AR(1), and the equation is

$$y_t = \beta_0 + \beta_1 y_{t-1} + \epsilon_t$$

In Figure 13.9 (page 13-18), we observed a positive, linear relationship between successive residuals from the exponential trend model fit to the DVD player sales data. Autocorrelation in the residuals from a regression model involving time-related variables indicates an AR(1) model might fit the data well.

We have two options for specifying an AR(1) model for the DVD player sales data. We can take the number of units sold per month to be the time series, or since the positive correlation in Figure 13.9 is for the residuals from the exponential trend model, we may want to take the logarithm of the number of units sold per month as the time series. In general, fitting an exponential model $y = ae^{bx}$ is the same as fitting a simple linear regression model with $\log(y)$ as the response variable instead of just y. Taking logarithms of both sides reveals a linear model based on $\log(y)$.

$$\log(y) = \log(ae^{bx}) = \log(a) + \log(e^{bx}) = \log(a) + bx$$

Figure 13.11 plots the number of units sold against the number of units sold lagged one period. Figure 13.12 plots the same points after taking logarithms. Figure 13.12 ($r = 0.95$) shows a stronger linear pattern than

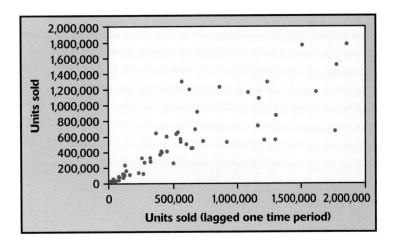

Figure 13.11 Lagged time series plot of the DVD player sales time series ($r = 0.80$).

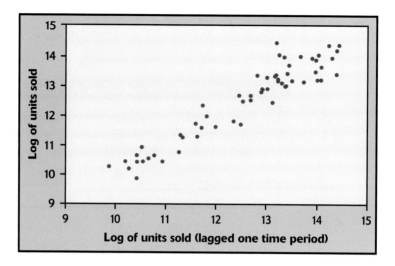

Figure 13.12 Lagged time series plot of the log of the DVD player sales time series ($r = 0.95$).

Figure 13.11 ($r = 0.80$) suggesting an AR(1) model will fit the DVD player sales data better after logarithms have been taken. We will proceed using the transformed time series in an AR(1) model.

EXAMPLE 13.9

CASE 13.1

DVD player sales

The AR(1) model for the time series $\log(y_t)$ where y_t is the number of DVD players sold in time period t is

$$\log(y_t) = \beta_0 + \beta_1 \log(y_{t-1}) + \epsilon_t$$

Statistical software calculates the fitted model to be

$$\widehat{\log(y_t)} = 0.43717 + 0.96486 \times \log(y_{t-1})$$

The model for DVD player sales in Example 13.9 looks like the simple linear regression models found in Chapter 10 except that the explanatory variable is the same as the response variable but lagged one time period. Despite the similarity in the form of the models, the details that made the least-squares line the best line in the setting of Chapter 10 are no longer valid with the autoregression model in Example 13.9. Most statistical software will estimate β_0 and β_1 in an autoregressive model using a method other than least-squares. The most common alternative to least-squares estimation *maximum* is called **maximum likelihood estimation** or simply MLE. The details of *likelihood* maximum likelihood estimation need not concern us. The general principle *estimation* of MLE is to use the parameter estimates that are most in agreement with our data. The agreement between parameter estimates and our data is measured in terms of "likelihood." Maximum likelihood estimates are the answer to the question "For what values of β_0 and β_1 are my data most likely to appear in a random sample?"

For the AR(1) model of Example 13.9, the MLE for β_1 is not too different from the least-squares estimate we would obtain by fitting a simple linear regression model to the time series. The least-squares estimate of β_1 is 0.94963 compared to the MLE of 0.96486. However, the intercept estimates differ more. The least-squares intercept estimate is 0.69 compared to the 0.43717 estimated via maximum likelihood. A difference of this magnitude will affect predictions you make with your model, so be sure to use software that was designed to estimate time series models correctly. Using simple linear regression to fit an AR(1) model is acceptable only if the time series has a mean near zero. The mean of the logarithm of the DVD player sales data is 12.5, so we need software that fits the AR(1) model using maximum likelihood estimation.

EXAMPLE 13.10

CASE 13.1

Forecasting July DVD player sales

Our DVD player sales time series ends with June 2002 sales of 1,617,098 units. June 2002 is the 63rd month of our time series, so our notation is $y_{63} = 1{,}617{,}098$. The AR(1) model of Example 13.9 relates DVD player sales for July 2002 to DVD player sales in June 2002, so we can forecast July's sales from the reported number of units sold in June.

The model fitted in Example 13.9 is for the time series $\log(y_t)$. This creates one extra step in our task of forecasting July 2002 DVD player sales. Our model will forecast $\log(y_{64})$ and we will need to calculate the forecast of y_{64} as the final step.

First, we use the model to forecast the logarithm of July 2002 DVD player sales.

$$
\begin{aligned}
\widehat{\log(y_{64})} &= 0.43717 + 0.96486 \times \log(y_{63}) \\
&= 0.43717 + 0.96486 \times \log(1{,}617{,}098) \\
&= 0.43717 + 0.96486 \times (14.2961437) \\
&= 14.23095
\end{aligned}
$$

Second, use your calculator's e^x button (or use software) to calculate the forecast of July 2002 DVD player sales.

$$
\begin{aligned}
\widehat{\log(y_{64})} &= 14.23095 \\
e^{\widehat{\log(y_{64})}} &= e^{14.23095} \\
\hat{y}_{64} &= 1{,}515{,}036
\end{aligned}
$$

Our AR(1) model for DVD player sales predicts sales of over 1.5 million units for July 2002.

In Case 13.1 (page 13-7), the exponential trend model predicted sales of 2.4 million units, and in Example 13.5 (page 13-14), the exponential trend-and-season model predicted sales of 2.1 million units. We noted that July sales are always below June sales (except in 1998), but that both these forecasts are greater than the June 2002 sales of 1.6 million. Our AR(1) model's forecast is an improvement in this respect because the forecast of 1.5 million units is below June 2002 sales of 1.6 million units.

Our forecast of July 2002 DVD player sales was based on an *observed* data value—June 2002 sales is a known value in our time series. If we wish to forecast August 2002 DVD player sales, we will have to base our forecast on an *estimated* value because July 2002 sales is not a known value in our time series.

EXAMPLE 13.11

CASE 13.1

Forecasting August DVD player sales

Following Example 13.10, we use the model to forecast the logarithm of August 2002 DVD player sales. Because our time series ends with y_{63}, the value of y_{64} is not known. In its place, we will use the value of $\hat{y}_{64}$ calculated in Example 13.10.

$$\widehat{\log(y_{65})} = 0.43717 + 0.96486 \times \log(\hat{y}_{64})$$
$$= 0.43717 + 0.96486 \times \log(1{,}515{,}036)$$
$$= 0.43717 + 0.96486 \times (14.23095)$$
$$= 14.16804$$

Finally, use your calculator's e^x button or software to calculate the forecast of August 2002 DVD player sales.

$$\widehat{\log(y_{65})} = 14.16804$$
$$e^{\widehat{\log(y_{65})}} = e^{14.16804}$$
$$\hat{y}_{65} = 1{,}422{,}662$$

We can continue in this manner to forecast DVD player sales for other future months.

In Chapters 10 and 11, we calculated prediction intervals for the response in our model. Associated with time series forecasts, your software may provide prediction intervals for future time periods. These intervals can be used and interpreted like the prediction intervals we saw in Chapters 10 and 11 although we will not present the details of their calculation.

The widths of time series prediction intervals generally increase as the time period of the prediction moves further beyond the end of the known time series. Predictions far into the future are subject to more uncertainty than predictions close to the end of the time series. Examples 13.10 and 13.11 demonstrate why this is true. Our forecast for July 2002 DVD player sales (y_{64}) depends on our estimated values of β_0 and β_1 and the *known* value of y_{63}. However, our forecast of y_{65} depends on estimated values of β_0, β_1, *and* y_{64}. The additional uncertainty involved in estimating y_{64} will make our prediction interval for y_{65} wider than our prediction interval for y_{64}. Statistical software provides the prediction intervals in the following table. The widths have been calculated—notice how the widths increase as we predict further into the future.

Time Period	Lower 95% PI	Upper 95% PI	Width
July 2002	678,324	3,383,804	2,705,480
August 2002	465,754	4,345,553	3,879,799
September 2002	349,161	5,133,999	4,784,838
October 2002	274,628	5,805,884	5,531,256
November 2002	223,067	6,384,087	6,161,020
December 2002	185,553	6,881,936	6,696,383
January 2003	157,263	7,309,011	7,151,748
February 2003	135,342	7,673,072	7,537,730
March 2003	117,982	7,980,825	7,862,843
April 2003	103,989	8,238,248	8,134,259
May 2003	92,539	8,450,757	8,358,218
June 2003	83,050	8,623,283	8,540,233

Autoregression models are useful when future values of a time series depend linearly on past values of the same time series. With software handling the calculation details, the concepts introduced in Chapters 10 and 11 for regression models apply equally well for autoregression models.

APPLY YOUR KNOWLEDGE

13.20 **Existing home sales.** The National Association of Realtors tracks monthly sales of existing homes in the United States. The file *ex13_020.dat* on the CD that accompanies this book has the existing home sales time series beginning in January 1968 and ending with July 2001. Use statistical software to analyze this time series.

(a) Make a time plot of the existing home sales time series.

(b) Describe any important features of the time series. Is there a strong, clear trend? If so, describe it. What about seasonal variation?

(c) Make a lagged time series plot (see Exercise 13.19 for an example) of the existing home sales time series. Does this plot suggest that an AR(1) model is appropriate for this time series? Why or why not?

13.21 **A model for existing home sales.** For this exercise, let y_t denote the number of existing homes sold (in thousands of units) during time period t. Use statistical software and the *ex13_020.dat* data file.

(a) Fit a simple linear regression model using y_t as the response variable and y_{t-1} as the explanatory variable. Record the estimated regression equation. Use this model to forecast August 2001 existing home sales.

(b) Fit an AR(1) model to the existing home sales time series. Record the estimated autoregression equation. Use this model to forecast August 2001 existing home sales.

(c) Compare the two estimated equations as well as the August 2001 forecasts from parts (a) and (b). State which of the two estimated models is preferred and briefly explain why.

Moving average models

The autoregression model AR(1) looks back *one* time period and uses that value of the time series in the forecast for the current time period. This works well when consecutive values of the time series are linearly related, but not all time series fit this pattern. With some time series, our forecasts can be improved by using the *average of many* past time periods. *Moving average* models use the average of the last several values of the time series to forecast the next value.

MOVING AVERAGE FORECAST MODEL

The **moving average forecast model** uses the average of the last k values of the time series as the forecast for time period t. The equation is

$$\hat{y}_t = \frac{y_{t-1} + y_{t-2} + \cdots + y_{t-k}}{k}$$

The number of preceding values included in the moving average is called the **span** of the moving average.

Some care should be taken in choosing the span k for a moving average forecast model. As a general rule, larger spans "smooth" the time series more than smaller spans by averaging many ups and downs in each calculation. Smaller spans tend to follow the ups and downs of the time series. With seasonal data, the length of the season is often used for the value of k.

EXAMPLE 13.12 **Moving averages for JCPenney sales**

Exercise 13.1 (page 13-5) asked you to plot and comment on quarterly JCPenney sales data. Figure 13.13 displays the JCPenney sales time series with the moving average forecast model based on a span of $k = 4$ overlaid. Note that the seasonal pattern in the time series is not present in the moving averages. Our moving averages are a smoothed version of our original time series. The moving averages follow the general movements of the time series but not every up and down.

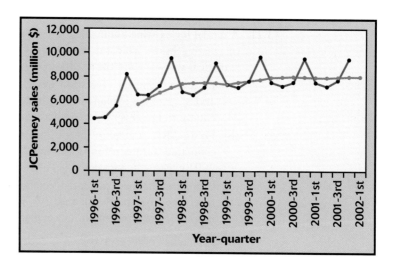

Figure 13.13 Moving averages based on a span of $k = 4$ (*purple*) overlaying the JCPenney quarterly sales data (*red*), for Example 13.12.

The JCPenney sales time series begins with the first quarter of 1996 and ends with the fourth quarter of 2001 for a total of 24 quarters. Figure 13.13 includes the forecasted value for the first quarter of 2002. The forecast is calculated as

$$\hat{y}_{25} = \frac{y_{24} + y_{23} + y_{22} + y_{21}}{4} = \frac{32,004}{4} = 8001$$

When a strong seasonal pattern is present, moving average models with the span set equal to the length of the season are similar to the trend-only models presented in Section 13.1. The moving averages will follow the long-run trend of the time series, but the ups and downs of the seasons are ignored. We would not expect the moving average model of Example 13.12 to forecast fourth-quarter JCPenney sales well.

EXAMPLE 13.13

Chicago Cubs attendance per game

Major League Baseball's Chicago Cubs have been playing their home games at Wrigley Field since 1916. The file *eg13_013.dat* on the CD that accompanies this book has the "attendance per home game" time series beginning in 1916 and ending in 2001.

Figure 13.14 displays the time series (red), moving averages based on a span of 5 years (purple), and moving averages based on a span of 20 years (black). The moving averages based on a span of 20 years highlight the general upward trend in attendance per home game at Wrigley Field without following the ups and downs in the data. The 5-year moving averages are not as smooth as the 20-year moving averages. The 5-year moving averages follow the larger ups and downs while smoothing the smaller changes in the time series.

The attendance time series has 86 observations ending with attendance per home game in 2001. Our interest is in forecasting attendance per home game for

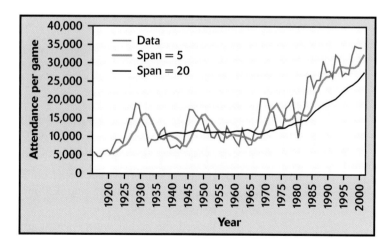

Figure 13.14 Chicago Cubs annual attendance per game time series (*red*) with 5-year (*purple*) and 20-year (*black*) moving averages overlaid, for Example 13.13.

2002. We will do this using both the 5-year moving averages and the 20-year moving averages.

$$\hat{y}_{87} = \frac{y_{86} + y_{85} + \cdots + y_{82}}{5} = \frac{162,522}{5} = 32,504$$

$$\hat{y}_{87} = \frac{y_{86} + y_{85} + \cdots + y_{67}}{20} = \frac{553,681}{20} = 27,684$$

The forecast based on the 20-year moving average is less than the forecast based on the 5-year moving average because the 20-year average includes the lower attendance figures of the 1980s. We might guess from looking at Figure 13.14 that the 5-year moving average forecast will be more accurate than the 20-year moving average forecast. In this case, actual 2002 attendance figures for Wrigley Field are now available. Total attendance was 2,693,096 for 81 home games, so the average attendance per home game was 2,693,096/81 or 33,248. The 5-year moving average forecast was more accurate.

——— ■ ■ ■

Once a new value becomes available (like 2002 attendance figures in Example 13.13), we update our time series to include the new observation. We can then calculate a forecast for the next time period (2003 attendance per home game in Example 13.13).

13.22 **Winter wheat prices.** The United States Department of Agriculture (USDA) tracks prices received by Montana farmers for winter wheat crops. The prices are tracked monthly in dollars per bushel. The file *ex13_022.dat* on the CD that accompanies this book has the wheat prices time series beginning in July 1929 and ending with October 2002 (880 months). Use statistical software to analyze this time series.

(a) Make a time plot of the wheat prices time series.

(b) Describe any important features of the time series. Be sure to comment on trend, seasonal patterns, and significant shifts in the series.

(c) Calculate 12-month moving averages and plot them on your time series plot from part (a).

(d) Calculate 120-month moving averages and plot them together with the time series from part (a) and the 12-month moving averages from part (c).

(e) Compare the 12-month and 120-month moving averages. Which features of the wheat prices time series does each capture? Which features does each smooth?

13.23 **Forecasting winter wheat prices.**

(a) Using the winter wheat prices from the previous exercise, calculate and compare the 12-month and 120-month moving average forecasts for November 2002.

(b) Record the actual winter wheat price received by Montana farmers for November 2002 from the Web page www.nass.usda.gov/mt/economic/prices/wwheatpr.htm. Which moving average forecast model provided the most accurate forecast for November 2002?

Exponential smoothing models

Moving average forecast models appeal to our intuition. Using the average of several of the most recent data values to forecast the next value of the time series is easy to understand conceptually. However, two criticisms can be made against moving average models. First, our forecast for the next time period ignores all but the last k observations in our data set. If you have 100 observations and use a span of $k = 5$, your forecast will not use 95% of your data! Second, the data values used in our forecast are all weighted equally. In many settings, the current value of a time series depends more on the most recent value and less on past values. We may improve our forecasts if we give the most recent values greater "weight" in our forecast

exponential smoothing calculation. *Exponential smoothing* models address both these criticisms.

There are several variations on the basic exponential smoothing model. We will look at the details of the *simple exponential smoothing model*, which we will refer to as, simply, the *exponential smoothing model*. More complex variations exist to handle time series with specific features, but the details of these models are beyond the scope of this chapter. We will only mention the scenarios for which these more complex models are appropriate.

EXPONENTIAL SMOOTHING MODEL

The **exponential smoothing model** uses a weighted average of the observed value y_{t-1} and the forecasted value $\widehat{y}_{t-1}$ as the forecast for time period t. The forecasting equation is

$$\widehat{y}_t = wy_{t-1} + (1 - w)\widehat{y}_{t-1}$$

The weight w is called the **smoothing constant** for the exponential smoothing model. The smoothing constant is a value between 0 and 1. Choosing w close to 1 puts more weight on the most recent value.

Choosing the smoothing constant w in the exponential smoothing model is similar to choosing the span k in the moving average model—both relate directly to the smoothness of the model. Smaller values of w correspond to greater smoothing of the ups and downs in the time series. Larger values of w put most of the weight on the most recent observed value, so the forecasts tend to follow the ups and downs of the series more closely.

EXAMPLE 13.14 **Philip Morris returns**

Table 1.10 (page 49) gives the monthly returns on Philip Morris stock for the period from June 1990 to July 2001 (134 months). Figure 13.15 displays the returns (red), the exponential smoothing model with $w = 0.5$ (purple), and the exponential smoothing model with $w = 0.1$ (black).

Using a smoothing constant of 0.5 puts more weight on the most recent values of the time series than does using a smoothing constant of 0.1. As a result, the

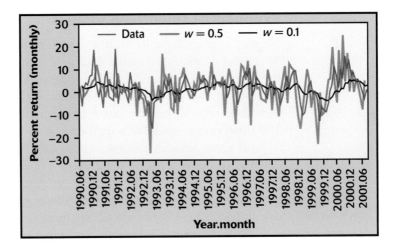

Figure 13.15 Philip Morris monthly stock returns (*red*) with two overlays, exponential smoothing model with $w = 0.5$ (*purple*) and exponential smoothing model with $w = 0.1$ (*black*), for Example 13.14.

purple curve tends to follow the ups and downs of the time series more closely than the smoother black curve. With a smoothing constant close to 0, the model will follow only the major changes in the time series.

A little algebra is needed to see that exponential smoothing models address the criticisms of moving average models. We will start with the forecasting equation for the exponential smoothing model and imagine forecasting the value of the time series for the time period $n + 1$ where n is the number of observed values in the time series. A specific example would be forecasting the August 2001 Philip Morris return y_{135} using the observed returns for the preceding $n = 134$ months.

$$
\begin{aligned}
\widehat{y}_{n+1} &= wy_n + (1 - w)\widehat{y}_n \\
&= wy_n + (1 - w)[wy_{n-1} + (1 - w)\widehat{y}_{n-1}] \\
&= wy_n + (1 - w)wy_{n-1} + (1 - w)^2\widehat{y}_{n-1} \\
&= wy_n + (1 - w)wy_{n-1} + (1 - w)^2[wy_{n-2} + (1 - w)\widehat{y}_{n-2}] \\
&= wy_n + (1 - w)wy_{n-1} + (1 - w)^2wy_{n-2} + (1 - w)^3\widehat{y}_{n-2} \\
&\;\;\vdots \\
&= wy_n + (1 - w)wy_{n-1} + (1 - w)^2wy_{n-2} + \cdots + (1 - w)^{n-2}wy_2 \\
&\quad + (1 - w)^{n-1}y_1
\end{aligned}
$$

Careful substitution and multiplication reveal a version of the forecast equation showing exactly how our forecast depends on the values of the time series. First, notice that the calculation of the forecast for y_{n+1} uses *all* available values of the time series $y_1, y_2, \ldots, y_n$, not just the most recent k values as a moving average model would. Second, the values are not equally weighted. Using a value of w close to 1 puts greater weight on the most recent observation. In this version of the forecast equation, we also see how the model gets its name. The coefficients in the forecast model decrease exponentially in value as you read the equation from left to right (with the exception of the last coefficient, $(1 - w)^{n-1}$). Exercise 13.25 has you explore what happens to the coefficients as you change the value of the smoothing constant w.

While the second version of our forecasting equation reveals some important properties of the exponential smoothing model, it is easier to use the first version of the equation for calculating forecasts.

EXAMPLE 13.15

Forecasting Philip Morris returns

Consider forecasting the Philip Morris August 2001 return using an exponential smoothing model with $w = 0.5$.

$$\hat{y}_{135} = 0.5 \times y_{134} + (1 - 0.5) \times \hat{y}_{134}$$

We need the forecasted value $\hat{y}_{134}$ to finish our calculation. However, to calculate $\hat{y}_{134}$ we will need the forecasted value of $\hat{y}_{133}$! In fact, this pattern continues, and we need to calculate all past forecasts before we can calculate $\hat{y}_{135}$. We will calculate the first few forecasts here and leave the remaining calculations for software. The forecast for the first time period $\hat{y}_1$ is always taken to be the actual value y_1 of the time series in the first time period.

$$\begin{aligned} \hat{y}_2 &= 0.5 \times y_1 + (1 - 0.5) \times \hat{y}_1 \\ &= (0.5)(3) + (0.5)(3) \\ &= 3 \end{aligned}$$

$$\begin{aligned} \hat{y}_3 &= 0.5 \times y_2 + (1 - 0.5) \times \hat{y}_2 \\ &= (0.5)(-5.7) + (0.5)(3) \\ &= -1.35 \end{aligned}$$

$$\begin{aligned} \hat{y}_4 &= 0.5 \times y_3 + (1 - 0.5) \times \hat{y}_3 \\ &= (0.5)(1.2) + (0.5)(-1.35) \\ &= -0.075 \end{aligned}$$

Software continues our calculations to arrive at a forecast for y_{134} of -3.812. We use this value to complete our forecast calculation for y_{135}.

$$\begin{aligned} \hat{y}_{135} &= 0.5 \times y_{134} + (1 - 0.5) \times \hat{y}_{134} \\ &= (0.5)(4.2) + (0.5)(-3.812) \\ &= 0.194 \end{aligned}$$

Our model forecasts a 0.194% return for Philip Morris stock in August 2001.

With the forecasted value for August 2001 from Example 13.15, calculating the forecast for September 2001 requires only one calculation.

$$\hat{y}_{136} = (0.5)(y_{135}) + (0.5)(0.194)$$

Once we observe the actual Philip Morris return for August 2001 y_{135}, we can enter that value into the forecast equation above. Updating forecasts from exponential smoothing models requires only that we keep track of last period's forecast and last period's observed value. In contrast, moving average models require that we keep track of the last k observed values of the time series. For this reason, exponential smoothing models are often preferred over moving average models especially if data storage is an issue.

The exponential smoothing model is best suited for forecasting time series with no strong trend or seasonal variation. Variations on the exponential smoothing model have been developed to handle time series with a trend (double exponential smoothing and Holt's exponential smoothing), with seasonality (seasonal exponential smoothing), and with both trend and seasonality (Winters' exponential smoothing). Your software may offer one or more of these smoothing models.

APPLY YOUR KNOWLEDGE

13.24 Philip Morris returns.

(a) Using the Philip Morris returns in Table 1.10 (page 49), calculate the first 4 forecasts $\hat{y}_1, \ldots, \hat{y}_4$ using an exponential smoothing model with $w = 0.1$. Do not use statistical software for these calculations.

(b) Now, use software to fit an exponential smoothing model with $w = 0.1$. Use the forecasts provided by the software to verify your hand calculations in part (a). Are your forecasts the same as those provided by your software?

(c) Provide a forecast for the August 2001 Philip Morris return based on your exponential smoothing model with $w = 0.1$ and compare this to the forecast calculated in Example 13.15.

(d) Write down the forecast equation for the September 2001 Philip Morris return based on the exponential smoothing model with $w = 0.1$.

13.25 It's exponential. Exponential smoothing models are so named because the coefficients

$$w, (1-w)w, (1-w)^2w, \ldots, (1-w)^{n-2}w$$

decrease in value exponentially. For this exercise, take $n = 11$. Use software to do the calculations.

(a) Calculate the coefficients for a smoothing constant of $w = 0.1$.

(b) Calculate the coefficients for a smoothing constant of $w = 0.5$.

(c) Calculate the coefficients for a smoothing constant of $w = 0.9$.

(d) Plot each set of coefficients from parts (a), (b), and (c). The coefficient values should be measured on the vertical axis while the horizontal axis can simply be numbered $1, 2, \ldots, 9, 10$ for the 10 coefficients from each part. Be sure to use a different plotting symbol and/or color to distinguish the three sets of coefficients and connect the points for each set. Also, label the plot so that it is clear which curve corresponds to each value of w used.

(e) Describe each curve in part (d). Which curve puts more weight on the most recent value of the time series when you are calculating a forecast?

(f) The coefficient of y_1 in the exponential smoothing model is $(1 - w)^{n-1}$. Calculate the coefficient of y_1 for each of the values of w in parts (a), (b), and (c). How do these values compare to the first 10 coefficients you calculated for each value of w? Which value of w puts the greatest weight on y_1 when you are calculating a forecast?

BEYOND THE BASICS: SPLINE FITS

smoothing spline

Modern computing capabilities have made possible many statistical tools that would otherwise be impossible or impractical. One such tool is the general purpose **smoothing spline**. A spline curve can be fit to any (x, y) data set. For a time series, we take x to be the time variable and y to be the time series values. A spline curve is a single curve consisting of a large number of polynomial curves pieced together in an optimal manner. The more polynomials used, the more flexible and less smooth the spline curve is.

The smoothness of a spline fit is determined by the choice of a positive-valued smoothing constant similar to the choice of w in the exponential smoothing model. The spline smoothing constant is often denoted by the lowercase Greek letter lambda (λ). Choosing a value of λ close to zero results in a very flexible and not very smooth spline curve. As greater values of λ are used, the spline curve becomes less flexible and more smooth.

Figure 13.16 displays three spline curves. The red spline curve offers the least amount of smoothing for the data ($\lambda = 0.000001$). With λ nearly 0, the spline curve passes through every point. The purple spline curve is based on $\lambda = 10,000$. This spline curve follows the ups and downs of the time series without trying to pass through each point. The black spline curve offers the

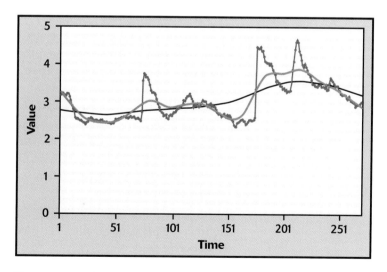

Figure 13.16 Spline curve models for three values of the smoothing constant λ: $\lambda = 0.000001$ (*red*), $\lambda = 10,000$ (*purple*), and $\lambda = 1,000,000$ (*black*).

highest degree of smoothing with $\lambda = 1,000,000$. The black spline curve reveals the overall pattern in the time series without following the smaller ups and downs.

Spline fits can help you quickly identify overall trends and seasonal variation as well as less regular patterns in your time series data. They are a powerful, modern exploratory data analysis tool.

SECTION 13.2 SUMMARY

■ Regression models relating a time series to a time variable are not always appropriate. Specifically, the residuals from such a model may exhibit high correlation. **Time series models** use past values of the time series to forecast future values of the time series.

■ The **first-order autoregression model,** AR(1), is appropriate when successive values of a time series are linearly related. The parameters of the model are often estimated using **maximum likelihood estimation** rather than least-squares estimation.

■ **Moving average forecast models** use the average of the last k observed values to forecast next period's value. k is called the **span** of the moving average. Larger values of k result in a smoother model.

■ The forecast equation for the **exponential smoothing** model is a weighted average of last period's observed value and last period's forecasted value. The degree of smoothing is determined by the choice of a smoothing constant w between 0 and 1. Values of w close to 0 result in a smoother model.

SECTION 13.2 EXERCISES

13.26 **A closer look at oranges.** Example 1.7 (page 20) looked at the trend and seasonal variation in the average monthly price of oranges. Figure 1.7 (page 21) is a time series plot of the data. The data is found in the file *fg01_007.dat* on the CD that accompanies this book.

(a) Make a lagged time series plot (see Exercise 13.19 for an example).

(b) Does the plot in part (a) suggest that an AR(1) model might be appropriate for the orange prices time series?

(c) Calculate the correlation between successive values of orange prices. Does the value of this correlation support your conclusion in part (b)? Why or why not?

13.27 **Least-squares or maximum likelihood?** Continue analyzing the orange prices data from the previous exercise. Let y_t denote the average price of oranges in time period t.

(a) Fit a simple linear regression model using y_t as the response variable and y_{t-1} as the explanatory variable. Record the estimated regression equation, the value of R^2, and the regression standard error s.

(b) Fit an AR(1) model to the orange prices time series. Record the estimated autoregression equation. Your software should also report a model R^2

and a model standard deviation (or model standard error)—record these values also.

(c) Compare the two estimated slopes and intercepts.

(d) Compare the model R^2-values and the model standard errors s. Do these values give any clear indication as to which fitting method (least-squares or maximum likelihood) is preferable?

13.28 Moving averages for orange prices.

(a) Calculate and plot (on a single time series plot) moving averages for spans of $k = 5, 10$, and 20.

(b) Comment on the smoothness of each moving average model in part (a). Which model would be best for forecasting monthly ups and downs in orange prices?

(c) Calculate and compare forecasts for January 2001 orange prices for each of the models in part (a). Which model provided the most accurate forecast? (The actual value of the orange prices time series for January 2001 is 224.2.)

13.29 Exponential smoothing for orange prices.

(a) Calculate and plot (on a single time series plot) exponential smoothing models using smoothing constants of $w = 0.1, 0.5$, and 0.9.

(b) Comment on the smoothness of each exponential smoothing model in part (a). Which model would be best for forecasting monthly ups and downs in orange prices?

(c) Calculate and compare forecasts for January 2001 orange prices for each of the models in part (a). Which model provided the most accurate forecast? (The actual value of the orange prices time series for January 2001 is 224.2.)

(d) Update your data by appending the January 2001 observed value of 224.2. Now forecast the February 2001 orange price with each of the models from part (a). Which model provided the most accurate forecast? (The actual value of the orange prices time series for February 2001 is 229.6.)

random walk models

13.30 A special AR(1) model. Random walk models for various financial time series are often mentioned in business literature. A simple random walk model specifies that one-period *differences* in the time series can be modeled as a constant term plus a random-error term. The equation for this random walk model is

$$y_t - y_{t-1} = \beta_0 + \epsilon$$

If we rewrite this equation solving for y_t, we get

$$y_t = \beta_0 + y_{t-1} + \epsilon$$

which is our AR(1) model with $\beta_1 = 1$. If we fit an AR(1) model and find that the β_1 estimate is close to 1, then a simple random walk model for the time series is another modeling option.

The CD that accompanies this book contains a data file named *ex13_030.dat*. Daily "USD to Euro" exchange rates beginning July 24, 2001 and ending July 23, 2002 are contained in the data file. The first exchange rate is 1.1509. This means that on July 24, 2001 a single U.S. dollar (USD) was worth 1.1509 euro (EUR).

    (a) Fit an AR(1) model to the exchange rate time series. Record the estimated forecast equation, the R^2-value, and the model standard error estimate. What evidence do you have that a simple random walk model for these exchange rates is appropriate?

    (b) Use the AR(1) forecast equation to predict the exchange rate for July 24, 2002.

    (c) For a simple random walk model, the estimate of β_0 is simply the average of all one period differences $y_t - y_{t-1}$ (call this average $\bar{y}_{\text{diff}}$) and the forecast equation is $\hat{y}_t = y_{t-1} + \bar{y}_{\text{diff}}$. Calculate the one-period differences and their average and use these to provide a "random walk forecast" for the exchange rate on July 24, 2002. Compare the random walk forecast to the AR(1) forecast.

    (d) You can get the actual exchange rate for July 24, 2002 at the Web site www.oanda.com/convert/fxhistory. Compare both forecasts to the actual exchange rate. Which model was more accurate?

13.31 Moving averages for exchange rates. Use statistical software to fit moving average models to the exchange rate data in the file *ex13_030.dat* on the CD that accompanies this book.

    (a) What would the moving average forecast equation be if we used a span of $k = 1$? (Your response to this part is not specific to the exchange rate data.)

    (b) Choose a value for k that is larger than 1 and less than 6 and calculate the corresponding moving averages. Be sure to report what value of k you chose.

    (c) Choose a value of k that is larger than 35 and less than 365 and calculate the corresponding moving averages. Be sure to report what value of k you chose.

    (d) Plot both sets of moving averages on a time series plot of the exchange rates. Be sure to label the plot clearly. Comment on the smoothness of both sets of moving averages.

13.32 Exchange rate moving averages continued.

    (a) Using a span of $k = 1$, forecast the USD to Euro exchange rate on July 24, 2002.

    (b) Use your model from part (b) of the previous exercise to forecast the USD to euro exchange rate on July 24, 2002.

    (c) Use your model from part (c) of the previous exercise to forecast the USD to euro exchange rate on July 24, 2002.

    (d) What would our moving average forecast equation be if we used a span of $k = 365$? What would this model forecast for July 24, 2002?

    (e) You can get the actual exchange rate for July 24, 2002 at the Web site www.oanda.com/convert/fxhistory. Compare these four forecasts to the actual exchange rate. Which model was closest to the actual exchange rate? Which model performed the worst on this forecast?

13.33 Exponential smoothing of exchange rates. Use statistical software to fit exponential smoothing models to the exchange rate data in the file *ex13_030.dat* on the CD that accompanies this book.

    (a) What would the exponential smoothing forecast equation be if we used a smoothing constant of $w = 1$? What about $w = 0$? (Your response to this part is not specific to the exchange rate data.)

(b) Choose a value for w such that $0 < w < 0.3$ and fit the corresponding exponential smoothing model. Be sure to report what value of w you chose.

(c) Choose a value of w such that $0.7 < w < 1$ and fit the corresponding exponential smoothing model. Be sure to report what value of w you chose.

(d) Plot both exponential smoothing models on a time series plot of the exchange rates. Be sure to label the plot clearly. Comment on the smoothness of each model.

(e) Using the models from parts (b) and (c), forecast the July 24, 2002 exchange rate and compare both these forecasts to the the actual exchange rate on that day. (You can get the actual exchange rate for July 24, 2002 at the Web site www.oanda.com/convert/fxhistory.)

STATISTICS IN SUMMARY

Standard regression methods like those described in Chapters 10 and 11 can be used with time series data with time as the explanatory variable and the time series as the response variable. These methods can help model both trend and seasonality in a time series although the models will generally not satisfy the regression assumptions needed for inference. Special time series models relate the current period's value to past values of the time series. These models are often referred to as "smoothing models." We rely on statistical software for fitting time series models and forecasting future values using these models. Here are the specific skills you should develop from studying this chapter.

A. TRENDS

1. Identify a long-run trend in a time series plot.
2. Use software to fit a regression model to the long-run trend.
3. Use a regression model to forecast future values of a time series based on a trend-only model.

B. SEASONS

1. Identify seasonal variation in a time series plot.
2. Model seasonal variation using indicator variables in a regression model.
3. Model seasonal variation using seasonality factors.
4. Forecast future values of a time series taking into account trend and seasonality.
5. Recognize seasonally adjusted data in a time series plot.

C. TIME SERIES MODELS

1. Identify autocorrelation in a lagged residual plot or a lagged time series plot.
2. Use software to calculate the autocorrelation of a time series.

3. Use software to fit a first-order autoregression model to a time series.

4. Forecast one or more time periods ahead using the software output from an autoregression model.

5. Use software to calculate the moving averages for a time series.

6. Understand how the choice of the span k of a moving average forecast model relates to the smoothness of the model.

7. Forecast next period's value using a moving average forecast model.

8. Use software to fit an exponential smoothing model to a time series.

9. Understand how the choice of the smoothing constant value w in an exponential smoothing model relates to the smoothness of the model.

10. Forecast next period's value using an exponential smoothing model.

CHAPTER 13 REVIEW EXERCISES

13.34 **Just use last month's figures!** Working with the financial analysts at your company, you discover that when it comes to forecasting various time series they often just use last period's value as the forecast for the current period. This strikes you as a very naive approach to forecasting.

(a) If you could pick the estimates of β_0 and β_1 in the AR(1) model, could you pick values such that the AR(1) forecast equation would provide the same forecasts your company's analysts use? If so, specify the values that accomplish this.

(b) What span k in a moving average forecast model would provide the same forecasts your company's analysts use?

(c) What smoothing constant w in a simple exponential smoothing model would provide the same forecasts your company's analysts use?

(d) Given your responses to parts (a), (b), and (c), are your company's analysts doing the best they can do for forecasting? Explain your response.

13.35 **Chicago Cubs attendance trend.** The file *eg13_013.dat* on the CD that accompanies this book contains the average attendance per home game beginning in 1916 and ending in 2001 (86 years). Use statistical software to fit trend models to this time series.

(a) Make a time series plot of the attendance per home game data. Is there a clear seasonal pattern in the time series?

(b) Fit a line to the data. Be sure to use attendance per home game as the response variable and year as the explanatory variable. Report the equation of the least-squares line, the R^2-value, and the regression standard error s.

(c) Now fit a second degree polynomial to the data. Report the equation of the quadratic model, the R^2-value, and the regression standard error s.

(d) Finally, fit a third degree polynomial to the data. Report the equation of the cubic model, the R^2-value, and the regression standard error s.

(e) Based on the R^2-values and the regression standard errors, which of your models from parts (b), (c), and (d) would you use as the trend equation for the Chicago Cubs average attendance per home game time series?

13.36 Forecasting the attendance trend. Refer to the previous exercise on Chicago Cubs attendance.

(a) Using the models from parts (b), (c), and (d) of the previous exercise, forecast average attendance per home game for the year 2002.

(b) The actual 2002 attendance per home game for the Chicago Cubs was 33,248. Which of the three models provided the most accurate forecast? Is this the same model you chose in part (e) of the previous exercise?

13.37 A seasonal model for DVD player sales. Following the method of Example 13.3 (page 13-10), fit a seasonal model for DVD player sales using indicator variables for the months. For the trend portion of the model, express the exponential trend fitted in Case 13.1 (page 13-7) as a linear trend with the logarithm of UNITS as the response variable. That is, use $\log(\text{UNITS}) = \beta_0 + \beta_1 x$ as the trend portion of the model.

(a) Write down the estimated trend-and-season model.

(b) Explain why no indicator variable is needed for December.

(c) Use this trend-and-season model to forecast July 2002 DVD player sales. Notice that your equation forecasts the *logarithm* of July 2002 DVD player sales, so you will need to use the e^x key on your calculator to complete the forecast. See Example 13.10 (page 13-25) for an example of such a calculation.

(d) Compare your forecast in part (c) to the forecast using seasonality factors in a trend-and-season model from Example 13.5 (page 13-14).

(e) Compare your forecast in part (c) to the AR(1) forecast from Example 13.10.

The file fg13_016.dat on the CD that accompanies this book contains the time series plotted in Figure 13.16 (page 13-35). The variable measured is the monthly average price in the United States for a pound of ground coffee. The time series begins on January 1980 and ends on May 2002 (269 months). The next five exercises refer to this data set.

13.38 Not regression. Use a time series plot of the coffee price data to help you explain why a trend-only or a trend-and-season model will not fit this time series well.

13.39 Moving average? Experiment with several different spans k and choose a value for k that results in a moving average model that smooths the minor ups and downs in the data but captures the major jump in the series that occurs about two-thirds of the way through the values. Use your chosen model to forecast the June 2002 average price for a pound of coffee. Provide a time series plot of the data with your model overlaid.

13.40 Exponential smoothing? Experiment with several different smoothing constants w and choose a value for w that results in an exponential smoothing model that smooths the minor ups and downs in the data but captures the major jump in the series that occurs about two-thirds of the way through the values. Use your chosen model to forecast the June 2002 average price for a pound of coffee. Provide a time series plot of the data with your model overlaid.

13.41 Why not autoregression? Fit an AR(1) model and plot it with the coffee price data. Can an AR(1) model "smooth the minor ups and downs in the data but capture the major jump in the series that occurs about two-thirds of the way through the values"? Use the AR(1) model to forecast the June 2002 average price for a pound of coffee.

13.42 **Check the forecasts.** The June 2002 average price for a pound of coffee is available now. Go to the Web site bls.gov/data/home.htm and click on "Create Customized Tables (one screen)" for the category "CPI—Average Price Data." On the window that opens next, choose "U.S. city average" and scroll down the list of items to select "Coffee, 100%, ground roast, all sizes, per lb. (453.6 gm)" and then click the "Get Data" button. A new window will open containing coffee price data. Verify you have the correct data by checking that the May 2002 price matches the last value in the file *fg13_016.dat*. Record the June 2002 average coffee price and compare your forecasts from the three previous exercises. Which model provides the most accurate forecast for June 2002?

Notes for Chapter 13

1. The UPS press release regarding the holiday shipping season was posted at www.ups.com/content/corp/about/news/art_4230.html.

2. The home page for the Monthly Retail Trade Survey is found at www.census.gov/mrts/www/mrtshist.html.

3. JCPenney quarterly net sales data were extracted from annual reports found at www.jcpenney.net/company/finance/finarch.html.

4. Consumer Electronics Association (CEA), www.ce.org. Data used by permission.

5. The number of Macintoshes shipped per fiscal quarter can be found in Apple's quarterly earnings press release. Follow the "Financial Releases" link at www.apple.com/find/sitemap.html to find the documents.

6. For example, details of the Durbin-Watson test for autocorrelation can be found in M. H. Kutner, C. J. Nachtschiem, W. Wasserman, and J. Neter, *Applied Linear Statistical Models,* 4th ed., McGraw-Hill, 1996.

Chapter 13

13.1 (a) Each year, sales are lowest for the first two quarters and then increase in the third and fourth quarters. Sales decrease from the fourth quarter of one year to the first quarter of the next. (b) The pattern described is obvious in the time plot. (c) There appears to be a positive trend, although the trend levels off after 1998. (d) The pattern described in part (a) is repeated year after year.

13.3 (a) Sales = $5903.22 + 118.75x$ with sales in millions of dollars and x takes on values 1, 2, ..., 24. (b) The fourth quarter of 1995. (c) The slope is the increase in sales (in millions of dollars) that occurs from one quarter to the next.

13.5 (a) Sales $= 7858.76 + 99.54x - 2274.21X1 - 2564.58X2 - 2022.79X3$. (b) If we know that $X1 = X2 = X3 = 0$, then we know that we are not in any of the first three quarters, and hence must be in the fourth quarter. (c) The intercept again represents the fourth quarter of 1995.

13.7 (a) The seasonality factors are 0.923, 0.885, 0.960, and 1.231 for quarters 1 through 4, respectively. (b) The average is 0.999, which is close to 1. The fourth-quarter seasonality factor of 1.231 tells us that fourth-quarter sales are typically 23.1% above the average for all four quarters. (c) The plot mimics the pattern of seasonal variation in the original series.

13.9 (a) Seasonally adjusted series is a little smoother. (b) Seasonally adjusting the DVD player sales data smoothed the time series a little but not to the degree that seasonally adjusting the sales data in Figure 13.7 did. This suggests that the seasonal pattern in the DVD player sales data is not as strong as it is in the monthly retail sales data.

13.11 (a) The dashed line in the time plot corresponds to the least-squares line. Using the trend-only model, we find that the sales for the first three quarters tend to be overpredicted and those for the fourth quarter are underpredicted. (b) The first quarter of 2002 is $8871.97 million. The fourth quarter of 2002 is $9228.22 million. (c) In the past, predictions in the first three quarters tended to be slightly more accurate than in the fourth quarter, so the first quarter of 2002.

13.13 (a) The first quarter of 2002 is $8188.83 million. The fourth quarter of 2002 is $11,359.94 million. (b) The first-quarter forecast has been multiplied by 0.923 as the trend-only model typically overpredicts the first quarter, while the fourth-quarter forecast has been multiplied by 1.231 to account for the fact that the trend-only model typically underpredicts the fourth quarter. (c) The trend-and-season model and the trend-only model with seasonality factors give similar predictions as both are adjusting the trend model for the seasonality effects.

13.15 (a) For the trend-only model $R^2 = 35\%$ and for the trend-and-season model $R^2 = 86.8\%$. The trend-and-season model explains much more of the variability in the JCPenney sales. (b) For the trend-only model, $s = 1170$, and for the trend-and-season model, $s = 566.7$. (c) The trend-and-season model closely follows the original series. (d) It is clear from the plot that the trend-and-season model is a substantial improvement over the trend-only model.

13.17 (a) The residuals for the fourth quarters are positive, and most of the residuals for the first three quarters are negative. (b) Autocorrelation is not apparent in the lagged residual plot. The correlation between successive residuals e_t and e_{t-1} is only 0.095.

13.19 (a) The other group of 10 points has December as the y coordinate. (b) The correlation of 0.9206 suggests a strong autocorrelation in much of the time series. (c) If we looked at the seasonally adjusted time series, the correlation would be closer to 0.9206. The outlying groups of points have the December sales as either the x or y coordinate, and this is what is reducing the correlation from 0.9206 to 0.4573. If there were a seasonal adjustment, these two sets of 10 points should no longer stand out from the remaining points.

13.21 (a) Fitting the simple linear regression model using y_t as the response variable and y_{t-1} as the predictor gives the equation $y_t = 34.0 + 0.992y_{t-1}$. The sales in July 2001 are 5170 thousands of units, so the forecast of August 2001 sales using this model is 5162.64. (b) Fitting the AR(1) model gives the equation $y_t = 13.9 + 0.996y_{t-1}$. The sales in July 2001 are 5170 thousands of units, so the forecast of August 2001 sales using this model is 5163.22. (c) The coefficients of y_{t-1} are very similar in parts (a) and (b), but the constants differ. The constant for the AR(1) model is much smaller than the constant (intercept) for the simple linear regression model. The August 2001 estimates are very close. The AR(1) model is preferred because it was estimated using maximum likelihood, which is preferred over least-squares for time series models.

13.23 (a) The 12-month moving average forecast predicts the November 2002 price by the average of the preceding 12 months. Averaging the last 12 values in the series gives the 12-month moving average forecast as $3.31. The 120-month moving average forecast predicts the November 2002 price by the average of the preceding 120 months. Averaging the last 120 values in the series gives the 120-month moving average forecast as $3.39. (b) Going to the Web page gives the actual winter wheat price received by Montana farmers for November 2002 as $4.28. The 120-month moving average forecast is slightly better, but neither captures the sharp rise in price that occurred over the latter part of 2002.

13.25 (a) When $w = 0.1$ the coefficients are 0.1, 0.09, 0.081, 0.0729, 0.06561, 0.059049, 0.0531441, 0.0478297, 0.0430467 and 0.0387420. (b) When $w = 0.5$ the coefficients are 0.5, 0.25, 0.125, 0.0625, 0.03125, 0.015625, 0.0078125, 0.0039062, 0.0019531 and 0.0009766. (c) When $w = 0.9$ the coefficients are 0.9, 0.09, 0.009, 0.0009, 0.00009, 0.000009, 0.0000009, 0.0000001, 0.0000000 and 0.0000000 (d), (e) The curves for $w = 0.9$ and $w = 0.5$ show a more rapid exponential decrease than the curve for $w = 0.1$ which decreases more slowly. The curve for $w = 0.9$ puts more weight on the most recent value of the time series in the computation of a forecast.

(f) The coefficients of y_1 are 0.348678, 0.000977, and 0.0000000 for the values $w = 0.1$, $w = 0.5$ and $w = 0.9$, respectively. The value of $w = 0.1$ puts the greatest weight on y_1 in the calculation of a forecast. Note that as indicated in the text, the values of the coefficients in the forecast model decrease exponentially in value with the exception of this last coefficient.

13.27 (a) Fitting the simple linear regression model using y_t as the response variable and y_{t-1} as the predictor gives $y_t = 100.4 + 0.572y_{t-1}$. $R^2 = 0.327$ and the regression standard error is 47.89. (b) An AR(1) model was fitted and the estimated autoregression equation is $y_t = 100.9 + 0.568y_{t-1}$. $R^2 = 0.327$ and the model standard error is 47.74. (c) The equations obtained for least squares and maximum likelihood are almost identical. (d) There is very little difference between the values of R^2 the model standard error s in this example. Thus there is no clear indication which fitting method is preferred, although in general maximum likelihood is preferred to least-squares in time series models.

13.29 (a), (b) The smoothness decreases as the value of w increases. The model with $w = 0.9$ would be best for forecasting the monthly ups and downs in orange prices. (c) The Minitab output for an exponential smoothing model provides forecasts for each value of the smoothing constant. The results are summarized below.

Smoothing constant w	Prediction
0.1	218.998
0.5	219.329
0.9	216.976

The forecasts are fairly similar, but the model with $w = 0.5$ provided the forecast that was closest to the actual value. (d) The January 2001 data point was added to the series and the models with the three weighting constants were fit to the new series. These were then used to forecast the orange price for February 2001. The results are summarized below.

Smoothing constant w	Prediction
0.1	219.518
0.5	221.764
0.9	223.478

The actual value in February was 229.6, so in this case the model with $w = 0.9$ provided the forecast that was closest to the actual value.

13.31 (a) If we use a span of $k = 1$, the moving average forecast equation would be $\hat{y}_t = y_{t-1}$. (b), (c), (d) We used $k = 4$ and $k = 100$. The moving averages with a span of $k = 4$ are not particularly smooth. The moving averages with a span of $k = 100$ are quite smooth.

13.33 (a) If we use an exponential smoothing forecast equation with smoothing constant of $w = 1$, the forecast equation is $\hat{y}_t = y_{t-1}$. If we use an exponential smoothing forecast equation with smoothing constant of $w = 0$, the forecast equation is $\hat{y}_t = y_{t-1}$. (b), (c), (d) We used $w = 0.2$ and $w = 0.9$.

The exponential smoothing model with $w = 0.2$ is smoother than the exponential smoothing model with $w = 0.9$ (e) Using the model from part (b), we find the forecast for the July 24, 2002 exchange rate to be 0.993532. Using the model from part (c), we find the forecast for the July 24, 2002 exchange rate to be 0.991149. The actual exchange rate on July 24, 2002 (from the web site www.oanda.com/convert/fxhistory) is 1.01140. Both forecasts underestimate the actual exchange rate.

13.35 (a) There are no clear seasonal patterns in the time series plot. (b) We fitted a line to the data using statistical software and obtained $\hat{y} = -487430 + 256.735(\text{year})$, $R^2 = 0.620$, and $s = 5048$. (c) We fitted a second degree polynomial to the data using statistical software and obtained $\hat{y} = 20082384 - 20752.3(\text{year}) + 5.36356(\text{year})^2$, $R^2 = 0.753$, and $s = 4091$. (d) We fitted a third degree polynomial to the data using statistical software and obtained $\hat{y} = -1101586137 + 1697733(\text{year}) - 872.171(\text{year})^2 + 0.149355(\text{year})^3$, $R^2 = 0.802$, and $s = 3684$. (e) Both the quadratic and cubic models appear to fit appreciably better than the straight line model. The cubic model fits a little better (a bit larger R^2 and a bit smaller s) than the quadratic model and might be a slightly better choice for the trend equation.

13.37 (a) To fit the trend-and-season model, we must first define 11 indicator variables, as in Example 13.3. We let Jan. $= 1$ if the month is January, 0 otherwise; Feb. $= 1$ if the month is February, 0 otherwise; and so on through November. Using software we obtained the estimated trend-and-season model $\ln(\text{UNITS}) = 10.691 + 0.069(\text{Case}) - 0.761(\text{Jan.}) - 0.786(\text{Feb.}) - 0.556(\text{Mar.}) - 0.544(\text{Apr.}) - 0.632(\text{May}) - 0.371(\text{Jun.}) - 0.525(\text{Jul.}) - 0.533(\text{Aug.}) - 0.041(\text{Sep.}) + 0.057(\text{Oct.}) - 0.225(\text{Nov.})$. (b) If Jan. $= 0$, Feb. $= 0$, Mar. $= 0$, ..., Nov. $= 0$, then we know the month must be December. Adding an indicator variable for December would be redundant. We can determine whether the month is December from the other indicator variables. (c) To use the equation in part (a) to forecast for July 2002, we notice that this would correspond to case 64. We set all indicators equal to 0 except the indicator for July, which we set to 1. We get $\ln(\text{UNITS}) = 14.582$ Thus, we forecast sales for July 2002 to be UNITS $= e^{14.582} = 2,152,198$. (d) The forecast using the seasonality factors in a trend-and-season model from Example 13.5 is UNITS $= 2,120,278$. The forecast in part (c) is slightly larger. (e) The forecast using the AR(1) model from Example 13.10 is 1,515,036. The forecast in part (c) is considerably larger.

13.39 We tried spans of 12, 24, and 36. The plot with span $k = 36$ does the best job of smoothing the minor ups and downs while still capturing the major jump about two-thirds of the way through the data. An even larger value of k might also work well. Using the moving average model with span $k = 36$, Minitab forecasts the June 2002 average price per pound of coffee as 3.25178.

13.41 The AR(1) model does not smooth the minor ups and downs in the data, although it does capture the major jump in the series that occurs two-thirds of the way through the values. Using the AR(1) model, Minitab forecasts the June 2002 average price per pound of coffee as 3.01245.

.